# WORKING IN MATH

by Christine Zuchora-Walske

www.12StoryLibrary.com



12-Story Library is an imprint of Bookstaves and Press Room Editions

Produced for 12-Story Library by Red Line Editorial

Photographs ©: ESB Professional/Shutterstock Images, cover, 1; Antoniooo/iStockphoto, 4; mTaira/Shutterstock Images, 5; photostock77/Shutterstock Images, 6; AP Images, 7; Sgt. Ken Scar/US Army, 8; Sean MacEntee CC2.0, 9; Avalon Studio/iStockphoto, 10; PhotoTalk/iStockphoto, 11; Solis Images/iStockphoto, 12; People Images/iStockphoto, 13, 21, 29; JSC/NASA, 14, 15, 28; The British Library, 16; Good Life Studio/iStockphoto, 17; Everett Historical/Shutterstock Images, 18; wk1003mike/Shutterstock Images, 19; zmeel/iStockphoto, 20; Lee Walters/iStockphoto, 22; Joseph Sohm/Shutterstock Images, 23; NOAA Legacy Photo/OAR/ERL/Wave Propagation Laboratory, 24; NWS/NOAA, 25; Seth Wenig/AP Images, 26; Barry Tuck/Shutterstock Images, 27

**Library of Congress Cataloging-in-Publication Data**
Names: Zuchora-Walske, Christine, author.
Title: Working in math / by Christine Zuchora-Walske.
Description: Mankato, MN : 12 Story Library, [2017] | Series: Career files | Audience: Grades 4 to 6. | Includes bibliographical references and index.
Identifiers: LCCN 2016047453 (print) | LCCN 2016051321 (ebook) | ISBN 9781632354464 (hardcover : alk. paper) | ISBN 9781632355133 (pbk. : alk. paper) | ISBN 9781621435655 (hosted e-book)
Subjects: LCSH: Mathematics--Vocational guidance--Juvenile literature.
Classification: LCC QA10.5 .Z83 2017 (print) | LCC QA10.5 (ebook) | DDC 510.23--dc23
LC record available at https://lccn.loc.gov/2016047453

Printed in the United States of America
022017

# Table of Contents

1

# Math Is Numbers, Shapes, Patterns, and Order

What is math, exactly? When people see the word *math*, they may just think of numbers. Or maybe they picture math problems, such as adding, subtracting, multiplying, and dividing. Numbers and calculations are a part of math. But math is also so much more.

Math is studying amounts and the concepts of more and less. It's understanding what numbers mean and represent. Math is studying shapes. It's understanding the structure of shapes, their size, and how they change over time. Math is studying patterns. It's recognizing order, sequence, and repetition.

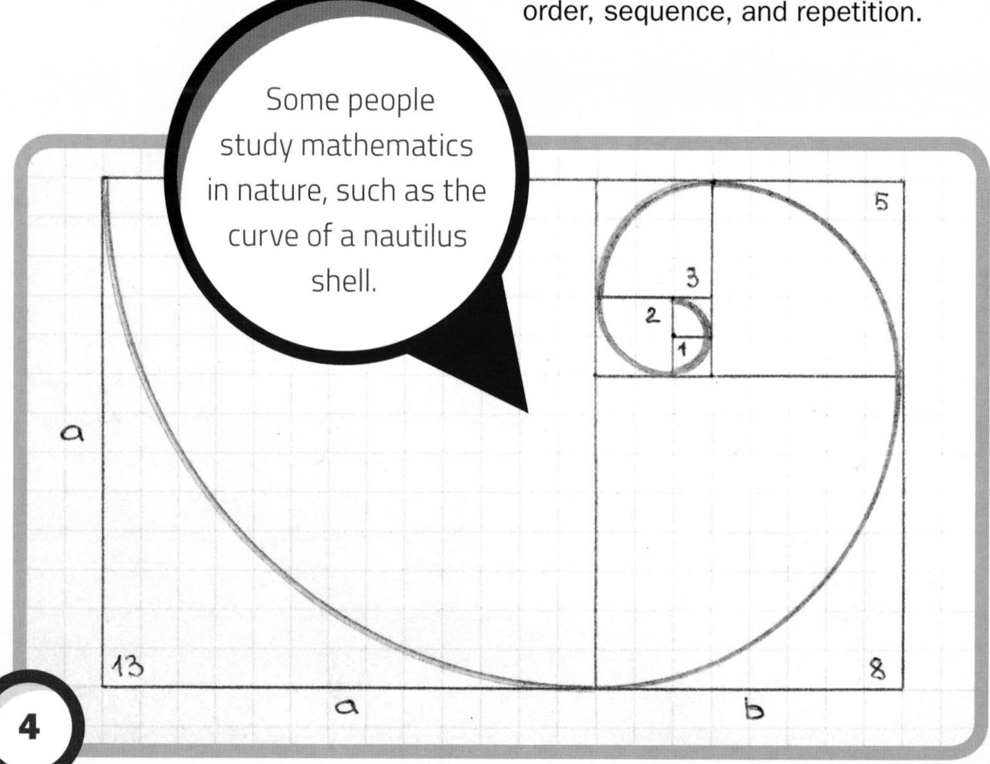


Some people study mathematics in nature, such as the curve of a nautilus shell.

In other words, math plays a role in nearly everything. The key to understanding how the universe works is math.

People use math in many ways, and math has several branches. A few of these branches are arithmetic, geometry, and applied math. Arithmetic is adding, subtracting, multiplying, and dividing numbers. Geometry is studying points, lines, angles, shapes, and solids. Applied math is using math to solve problems in any other science.

**35,000**

**Age in years of the oldest known math tool, a measuring stick from Swaziland.**

- Math is studying numbers, shapes, patterns, order, and change.
- To understand the universe, people need to understand math.
- Arithmetic, geometry, and applied math are three of the branches in mathematics.

Every career uses math in some way. Some careers use more math than others. People also use math in their everyday lives.

A baseball player uses math when figuring out the right trajectory for a pitch.

# 2 Math Takes Curiosity, Patience, and Hard Work

Many people think math is just calculating. They think it is a set of rules and formulas in arithmetic, algebra, geometry, and so on. They picture mathematicians as people who are really good at calculating.

But math is about more than finding the one right answer to an equation. Today's careers in math have much more to do with creative problem solving than number crunching. Mathematicians spend time thinking about how to transform information in the world into numbers. To prepare for a math-related career, it's important to learn as much math as possible.

Math is hard—even for math experts. So what does it take to succeed in math? Most experts agree that it's important to be persistent. It's also important to ask questions, even if the answer is hard to find. For example, a famous computer science problem about math is called the P versus

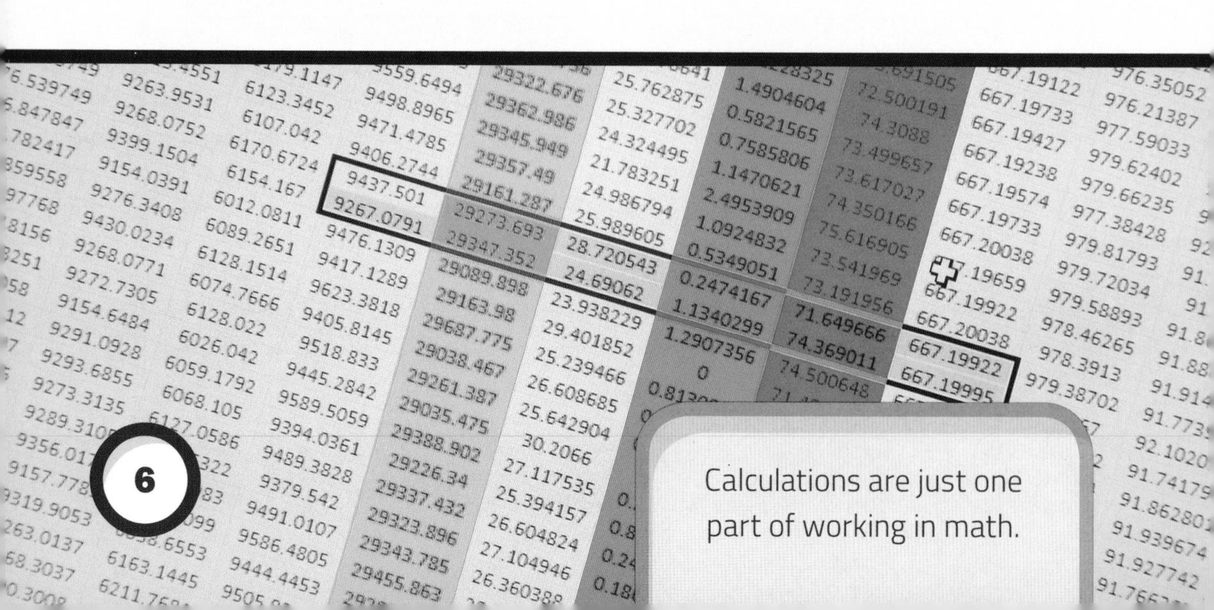

Calculations are just one part of working in math.

## EDUCATION

Most math-related jobs need some college education in math or a math-related science. For example, animators and cartographers can have associate's degrees, which take two years to earn. Architects, market researchers, air traffic controllers, and meteorologists typically have a bachelor's degree. Those degrees usually take four years to earn. Mathematicians frequently go back to school to earn additional college degrees, such as a master's or a doctorate.

NP problem. Mathematicians first thought about this question in 1956. The problem remains unsolved today, but mathematicians continue to work at it.

Albert Einstein spent many years searching for a unified theory of electromagnetism and gravity but never succeeded.

**100**

**Percentage of the 25 top-paying US jobs in 2016 that were math-related careers.**

- Math is about more than finding one right answer to an equation.
- Today's mathematicians have to be creative problem solvers.
- Math success takes curiosity, an open mind, and determination.

# 3

# Air Traffic Controllers Keep Planes from Crashing

People may not think about math when they picture an air traffic controller (ATC). But that job uses a lot of math.

ATCs are the traffic cops of the sky. They direct the movements of aircraft. About 5,000 aircraft are flying over the United States at any moment. ATCs guide all of them on the ground, during takeoff, through the sky, and during landing. ATCs make sure pilots follow their flight plans and stay on time. They use math to make sure each plane's arrival and departure takes a certain amount of time.

ATCs for the US Army undergo 15 weeks of on-the-job training.

# 50,000

**Approximate number of flights ATCs in the United States handle each day.**

- ATCs direct the movements of aircraft to keep them from crashing.
- Pilots get information and help from ATCs during emergencies.
- ATCs must do a lot of mental math and quick thinking.
- ATCs must have good memory, teamwork, and communication skills.

## THINK ABOUT IT

In the United States, ATCs are required to have nine hours of rest between shifts. Why do you think that's necessary?

Thousands of aircraft can soar through the air because ATCs keep them from crashing into one another. ATCs use math to tell how far apart planes are. They calculate a plane's velocity by determining its speed and direction. ATCs tell pilots about weather and ground conditions. For example, strong winds can change a plane's direction. ATCs use math to help planes stay on track. They also tell pilots what to do in emergencies.

ATCs must understand and act on lots of information quickly and constantly. They have to do fast math in their heads to understand distance, speed, and altitude. They need to manage several aircraft at once. They have to make quick decisions. Teamwork and clear communication are vital.

ATCs sometimes work in towers such as this one.

4

# Animators Bring Art to Life

Animation is everywhere. It's in apps, video games, and websites. It's in ads, TV shows, movies, and videos.

Animators create art that looks as though it is moving. But animators are not just artists. They need to know a lot of math to bring a still picture to life. The characters, objects, and backgrounds in pictures are made up of shapes. The shapes differ in type and size. Animators use geometry to understand the structure of a picture.

To make a picture look as though it is moving, an animator must create a sequence of pictures. Each picture in the sequence is slightly different from the one before it. The animator uses geometry and algebra to figure out how the shapes in the picture must change to show movement.

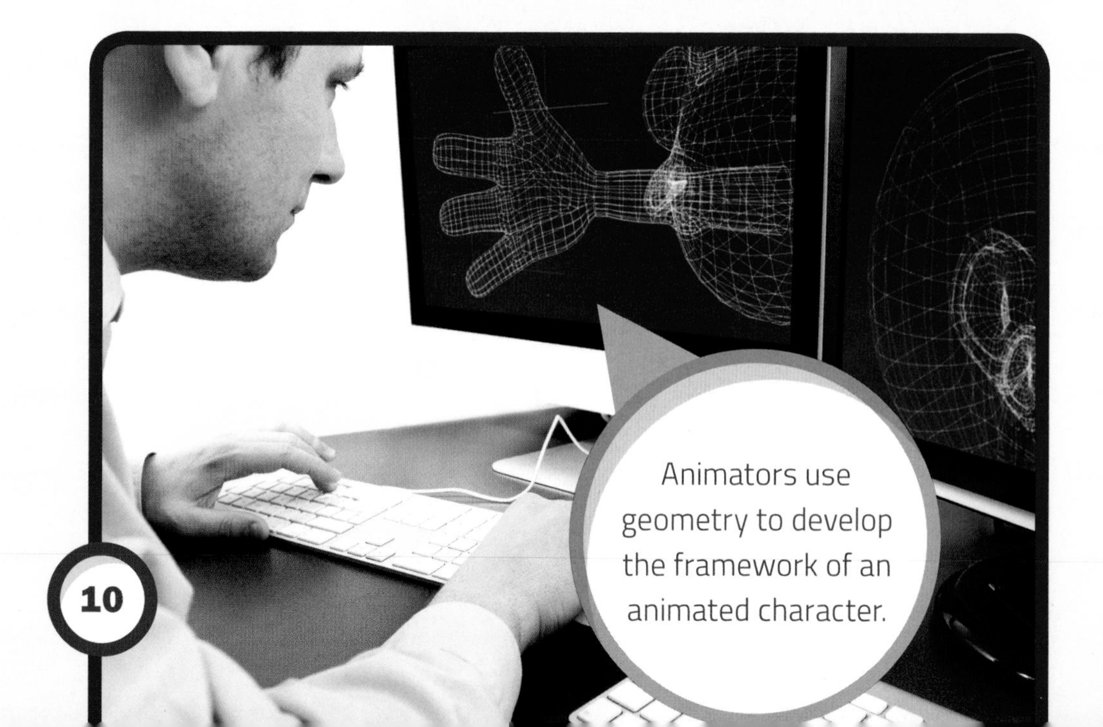

Animators use geometry to develop the framework of an animated character.

Animators also use a type of math called calculus. It helps them figure out how quickly the length, area, and volume of an object should change. The animators for *Finding Nemo* and *Finding Dory* used calculus to figure out how to show light underwater.

Animators draw pictures by hand or with computers. They usually use computer programs to animate their images. But animators must still understand the math behind shapes and motion.

Animators in *Finding Nemo* had to use math to figure out how light bends underwater.

**4**

**Average amount of screen time, in seconds, an animator creates in one week.**

- Animators make art that seems to move.
- Geometry helps animators understand the structure of pictures.
- Algebra and calculus help animators create the illusion of motion.
- Animators usually use computer animation programs.

## FRAME RATE

In animation, the term *frame rate* means the number of still pictures, or frames, shown per second. Most professional videos show at least 24 frames per second (fps). A video with a lower fps looks choppy or jerky. A video with a higher fps looks very realistic.

# 5

# Architects Design Beautiful, Safe, and Useful Buildings

Architects are people who design buildings. They find out what people need and what they can afford. Then they try to create a building that is as safe, useful, and pleasant as possible.

Architects have to explain their ideas in a plan. A plan uses words, numbers, drawings, and models. A plan helps other people understand the architect's ideas. Builders can use the plan as a construction guide.

Many architects use computer programs when drawing up plans.

## THINK ABOUT IT

Architects need to be good at math. But they need to have other skills, too. What else do architects need to know to create spaces that please people?

## 20

**Percentage of architects who are self-employed.**

- Architects find out what people need in a building and can afford.
- They design buildings that are as safe, useful, and pleasant as possible.
- Architects express their ideas in plans, using words, numbers, drawings, and models.
- Architects use math to draw up their plans.

Architects use geometry, algebra, and calculus when making their plans. For example, an architect has to know how much weight a certain type of wall can hold. Geometry also helps architects figure out how to fit all the pieces of a building together in the available space. Arithmetic helps architects scale their building-size ideas into desktop-size plans.

Architects use special rulers to get their plans just right.

6

# Astronauts Explore and Survive in Space

Astronauts command, fly, or serve as crew on spacecraft. Some astronauts have traveled to the moon. But most human space travel happens while orbiting Earth. Today, that means working on the International Space Station (ISS). The ISS is a research station where astronauts experience near weightlessness.

A spacecraft approached the ISS with supplies.

Astronauts do many jobs on the ISS. They work on science experiments. They study their own bodies. They learn how gravity affects humans over time. They keep the ISS running properly. They constantly check, clean, tune up, and fix its parts and machines.

ISS work uses a lot of math. For example, when using the ISS's robotic arm, astronauts need to do math in their heads. They can't let go of the hand controls. Even docking at the ISS takes some tricky math. To speed up, an astronaut must slow the spacecraft a little, so it fights the pull of Earth's gravity less. Then gravity pulls more and makes the craft go faster.

Astronauts aboard the ISS need to use math in other ways, too.

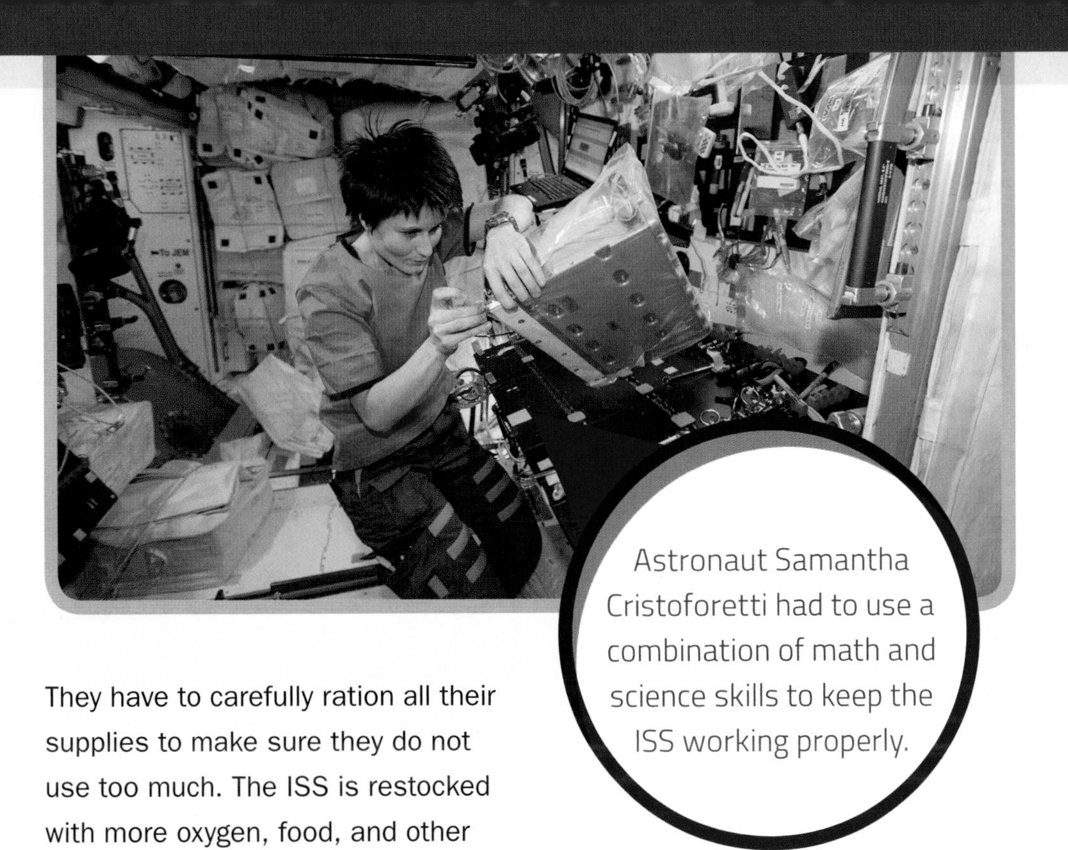

Astronaut Samantha Cristoforetti had to use a combination of math and science skills to keep the ISS working properly.

They have to carefully ration all their supplies to make sure they do not use too much. The ISS is restocked with more oxygen, food, and other supplies every few months.

**2**

**Years astronaut candidates train before being selected to work on a flight.**

- Astronauts command, fly, or serve as crew on spacecraft.
- Most human space travel today happens aboard the ISS.
- Astronauts use math in their scientific and maintenance work.

## MATH SAVES LIVES

Apollo 13 launched for the moon in April 1970. An oxygen tank exploded, badly damaging the craft. Its three astronauts had to ride home in the tiny moon lander. The lander was attached to the main craft. Commander James Lovell figured out how to get home without using the main craft's computer. With a pencil, he scribbled calculations in a moon lander manual. That math got the men home safely.

# 7

# Cartographers Make Maps

A cartographer is someone who makes maps. A map is any picture or model that shows information about a place. But maps show more than just the shape of the land, the names of places, and the boundaries between them. Maps can be crammed with fascinating information. Cartographers use creativity and logic to figure out the best way to show that information.

To gather information, cartographers may visit certain places and observe. Or they may study data collected by others. The information they collect can be anything from distances and heights to population and rainfall. Cartographers may study aerial or satellite photos, too.

In the past, all maps were hand-drawn or printed. This is called paper cartography. People still create maps

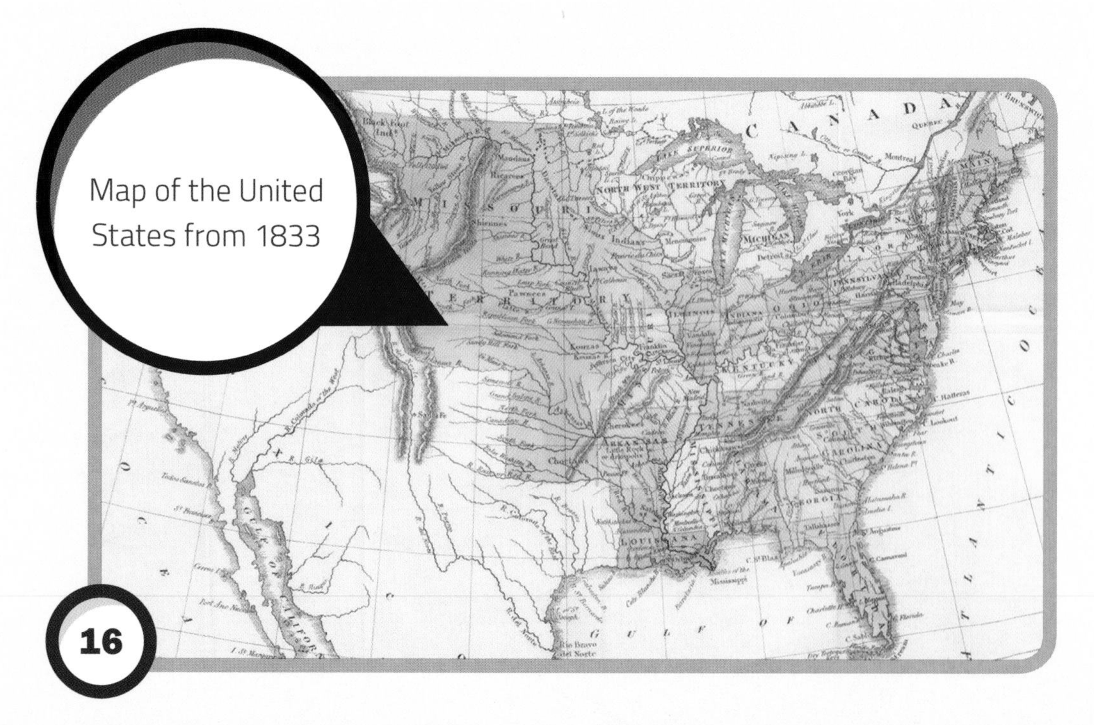


Map of the United States from 1833

this way. Paper maps use scales to correctly represent the relative distance between places.

Today, cartographers can also use a computer to create maps. This is called a geographic information system (GIS). GIS is any computer that gathers, stores, checks, or displays information about places. In GIS, people can choose which data to show. For example, a scientist could use GIS to show animal migration or polar ice cover over time.

## 29

**Expected percent increase in the number of cartography jobs available by 2024.**

- A cartographer is a mapmaker.
- A map is any picture or model that shows information about a place.
- In the past, maps were created on paper.
- In GIS, people choose which data to show and can show it in different ways.

Today, many people rely on Google Maps for navigation.

8

# Cryptologists Make and Break Codes

People, businesses, and governments all have information they want to keep private. For example, people may want to protect information about their personal lives. Businesses may want to protect their finances. Governments may want to protect military secrets. Much of this information is stored in computers.

The best way to hide information is to change it into a code. Only people who know the code can read the information. Cryptology is the science of making and breaking codes. Cryptology has two main parts. Cryptography is the process of encrypting information, or changing it into a code. Cryptanalysis is decrypting information, or breaking a code and making information readable.

The German Enigma machine used during World War II had 150 quintillion possible encryptions for messages.

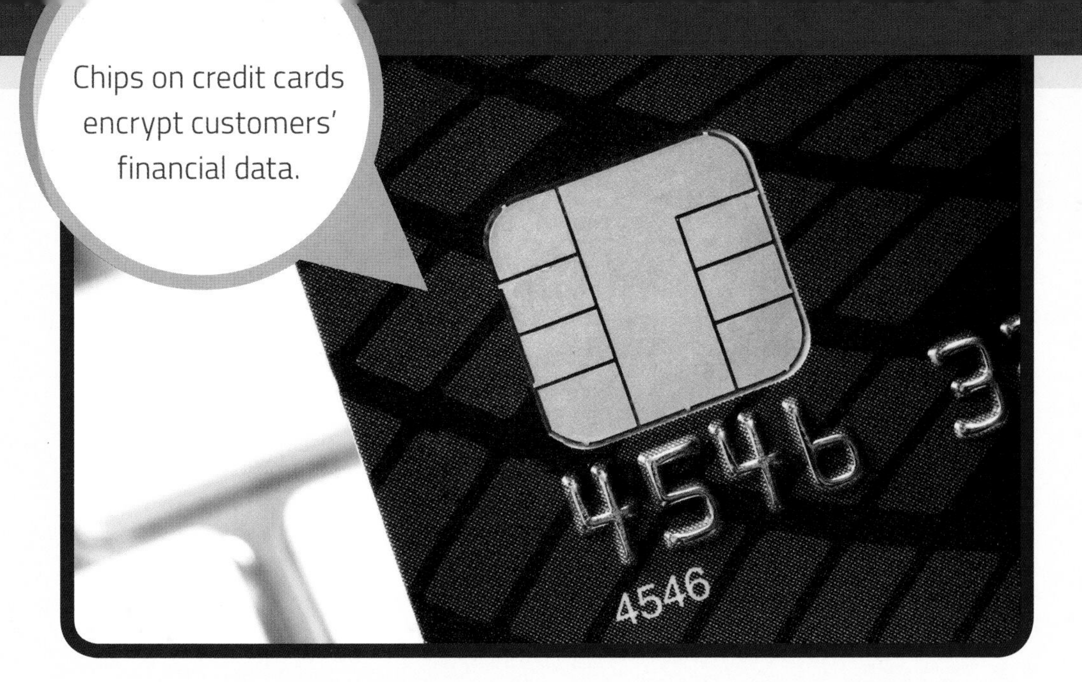


Chips on credit cards encrypt customers' financial data.

Cryptologists use math in both making and breaking codes. To encrypt information, they create a key. The key is an algorithm, or a complex set of steps that must be followed. The algorithm is based on a math problem that is too hard for a human to solve. Cryptologists may also decrypt secret messages sent by others. They might do this to help government, military, or law enforcement agencies. They use their knowledge of encryption and advanced math to crack the code.

**$76,470**

**Average annual salary earned by a cryptanalyst in the United States.**

- Cryptology is the science of making and breaking codes.
- Cryptography is encrypting information, or changing it into a code.
- Cryptanalysis is decrypting information, or breaking a code.
- Cryptologists use math in both making and breaking codes.

## THINK ABOUT IT

A cryptanalyst sometimes breaks codes to try to find out others' secrets. When is it okay to do this?

9

# Forensic Scientists Use Math to Solve Crime Puzzles

Forensic scientists are problem solvers. When crimes occur, forensic scientists find the facts. They gather evidence and study it. They use science and math to figure out what really happened.

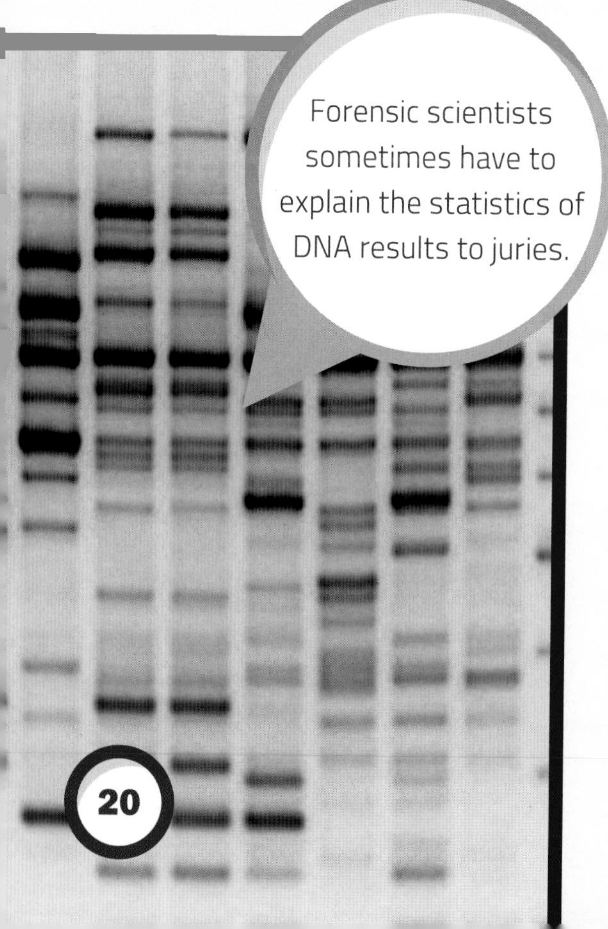

Forensic scientists sometimes have to explain the statistics of DNA results to juries.

Forensic scientists collect evidence by observing the scene, writing notes, drawing sketches, and taking photos. They may gather samples, such as fingerprints, blood, and hair. To evaluate the evidence, scientists use a lot of math. For example, DNA tests show the likelihood that two samples belong to the same person. This is expressed as a percentage. Scientists use 13 specific points in DNA sequencing for comparison. These are called genetic markers. The likelihood of two unrelated people having 1 out of 13 genetic markers match is 7.5 percent.

Forensic scientists also use math when helping police figure out how a person was injured or killed. The crime scene might include bloodstains. Scientists measure where the blood landed and the size and shape of the bloodstains.

This information can help show how hard the victim was hit, stabbed, or shot. A gunshot would spread many small drops of blood over a wide area. Being hit or stabbed would form larger drops of blood in a smaller area, close to the attack.

**100**

**Minimum number of years forensic scientists have been using fingerprints to solve crimes.**

- Forensic scientists help solve legal questions.
- They gather evidence by observing, drawing, writing notes, taking photos, and collecting samples.
- Forensic scientists measure and test evidence to figure out what happened.

Forensic scientists have to be precise in their measurements.

10

# Market Researchers Help Businesses Find Customers

Market researchers study what people think. They try to find out what products and services people want. They discover what people are looking for in these products and services. They also try to figure out how much people will pay and who the most likely customers are.

For example, market researchers at Target study the purchases made by their customers. They send customers who buy swimsuits in April coupons for sunscreen in July. Market researchers at Target have even been able to tell when a woman is pregnant based on her purchases.

Market researchers use math in most of their tasks. They gather and record data. They study past sales of products or services. They look for patterns in the data. They use this information to predict

Market researchers have helped Target expand to every state except Vermont.

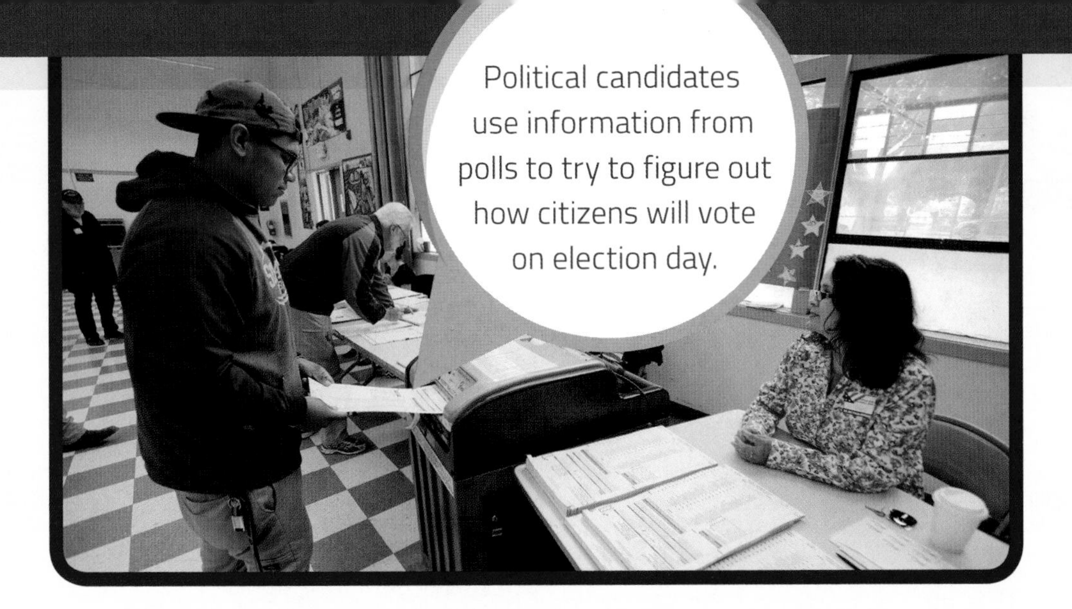

Political candidates use information from polls to try to figure out how citizens will vote on election day.

future sales. This process is called forecasting.

They also study the ways competing companies do business. They examine prices and sales. They study how different companies advertise and deliver their products and services. They use all this information to make suggestions for future business decisions.

**500,000**

**Approximate number of market researchers in the United States.**

- Market researchers discover what products and services people want.
- They also discover how much people will pay and who are the most likely customers.
- Market researchers use math to gather and study data.
- They use data to suggest future business decisions.

## POLITICAL POLLS

Political polling is a kind of market research. Pollsters ask people for their opinions on people who are running for elected offices. They also ask for opinions on issues that affect the public. They try to figure out whom and what voters will probably support. Pollsters may work for a candidate or a cause. Or they may simply be trying to predict voting results.

11

# Meteorologists Predict the Future

Meteorologists, or atmospheric scientists, study Earth's atmosphere, weather, and climate. They look at evidence from the past. They observe weather happening right now. They predict future weather and climate. They think about how weather and climate affect Earth and all the things that live on it.

Meteorologists use math in almost every part of their job. They measure temperature, precipitation, wind speed, humidity, and many other conditions in the atmosphere. They record all this information.

Meteorologists study weather and climate from the past to look for patterns. They use

Meteorologists use math when categorizing tornadoes.

## ANCIENT CLIMATE

How can meteorologists study the climate of the ancient past, before people started keeping track of weather? Clues about Earth's long-ago atmosphere exist in tree rings, coral reefs, polar ice, and the mud at the bottoms of lakes and oceans. These typically form in yearly layers. Scientists can study the chemicals trapped in them to learn about Earth's past atmosphere.

these patterns to predict what will happen in the future. They forecast the weather. Meteorologists also help people understand changes to the weather, such as air pollution and droughts. For example, they compare annual precipitation levels to past averages to show how severe a drought is.

**100**

**Maximum distance, in miles (161 km), that US meteorologists were off when forecasting the landfall of hurricanes in 2012.**

- Meteorologists are also called atmospheric scientists.
- They study past and present atmosphere, weather, and climate.
- Meteorologists use math to measure and analyze past and present conditions.
- They also use math to predict future conditions.

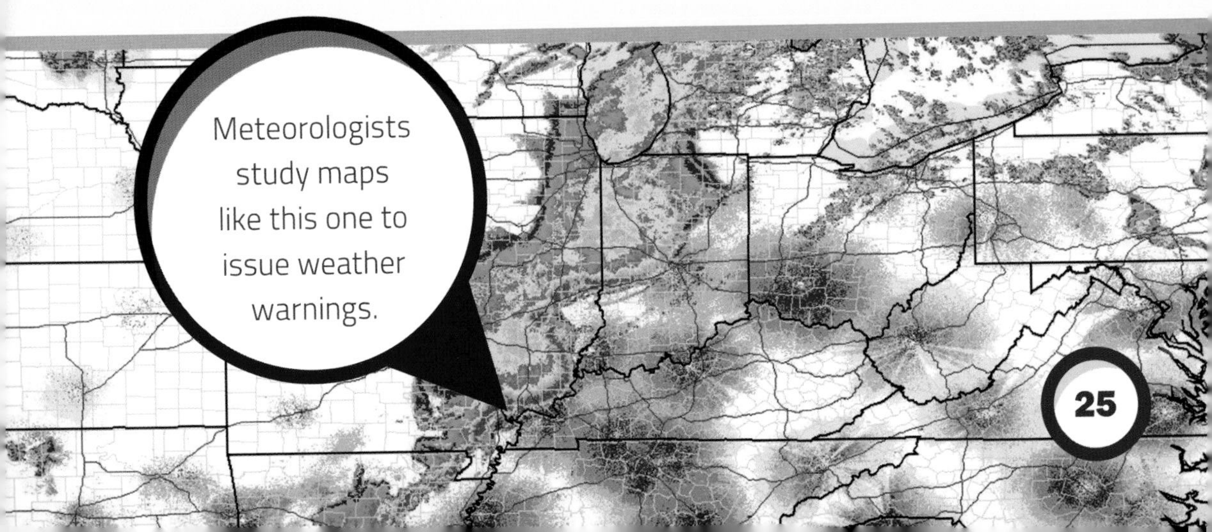

Meteorologists study maps like this one to issue weather warnings.

12

# Computers Encourage Math Creativity and Cooperation

Computers are changing the way people do math. They are changing math careers and math itself. Today's math workers still need to understand rules and formulas. But they do not need to do much calculating. Computers are much faster and better than humans at that. Instead, math careers are shifting to focus on transforming questions about the world into math language. Mathematicians must

IBM built a supercomputer called Watson, which easily beat *Jeopardy!* champions Ken Jennings and Brad Rutter in 2011.

reason *about* numbers—not just *with* numbers.

Math still needs human brains, though. Only people can create and ponder abstract math ideas. Those are ideas that are not about specific numbers. Rather, they are about how math itself works.

Computers make it easier for mathematicians to work together on abstract ideas. The Internet lets them cooperate, even when they are far apart. Many people can participate at the same time. They can share both successful and failed ideas. Online cooperation is speeding up the process of asking and answering abstract math questions.

# 93,000 trillion

**Number of calculations per second the world's fastest computer could do in 2016.**

- Computers are better and faster at calculating than humans are.
- Humans still need to figure out abstract math ideas.
- The Internet makes it easier for mathematicians to work together.

Scientists and physicists depend on supercomputers at the European Organization for Nuclear Research (CERN) in Geneva, Switzerland.

# Other Jobs to Consider

## Actuary

Description: Use math and statistics to evaluate risks and rewards for companies, especially in insurance
Training/Education: Bachelor's degree
Outlook: Growing
Average salary: $97,070

## Cost Estimator

Description: Estimate the amount of time, money, and resources a project will take
Training/Education: Bachelor's degree
Outlook: Growing
Average salary: $60,390

## Pharmacy Technician

Description: Calculate dosages to fill patients' medication prescriptions written by doctors
Training/Education: High school diploma
Outlook: Growing
Average salary: $30,410

## Project Manager

Description: Plan and oversee the budget, schedule, and other details of a specific project
Training/Education: Bachelor's degree
Outlook: Growing
Average salary: $90,000

# Glossary

**abstract**
Based on an idea rather than something that can be seen or touched.

**altitude**
Height above sea level.

**atmosphere**
The entire layer of air surrounding Earth.

**calculate**
To find by adding, subtracting, multiplying, dividing, or a combination of these processes.

**data**
Facts or information.

**evidence**
An outward sign of something's existence or occurrence.

**formula**
In math, a rule stated with math symbols.

**gravity**
The force that pulls objects toward Earth's center.

**logic**
A thought process that uses the relationships between facts, such as cause and effect, to reach a final idea or decision.

**reason**
To think logically.

# For More Information

## Books

Gunzenhauser, Kelly. *Find Your Future in Mathematics.* Ann Arbor, MI: Cherry Lake Publishing, 2017.

Simons, Lisa M. Bolt. *Unusual and Awesome Jobs Using Math: Stunt Coordinator, Cryptologist, and More.* North Mankato, MN: Capstone Press, 2015.

Zuchora-Walske, Christine. *Mathematics in the Real World.* Minneapolis, MN: Abdo Publishing, 2016.

# Index



About the Author

Christine Zuchora-Walske has been writing and editing children's books and magazines for 25 years. She writes about science, history, and current events. Christine lives in Minneapolis with her husband and two children.